The Earth

This is a map of the earth.
You can see that part of the earth is land.
You can see that part of the earth is covered with water.

Color the land you see green.
Color the water you see blue.

Note: Use this page along with the poster provided in this unit.

Land and Water

There are many shapes and sizes of land forms.

There are many shapes and sizes of bodies of water too.

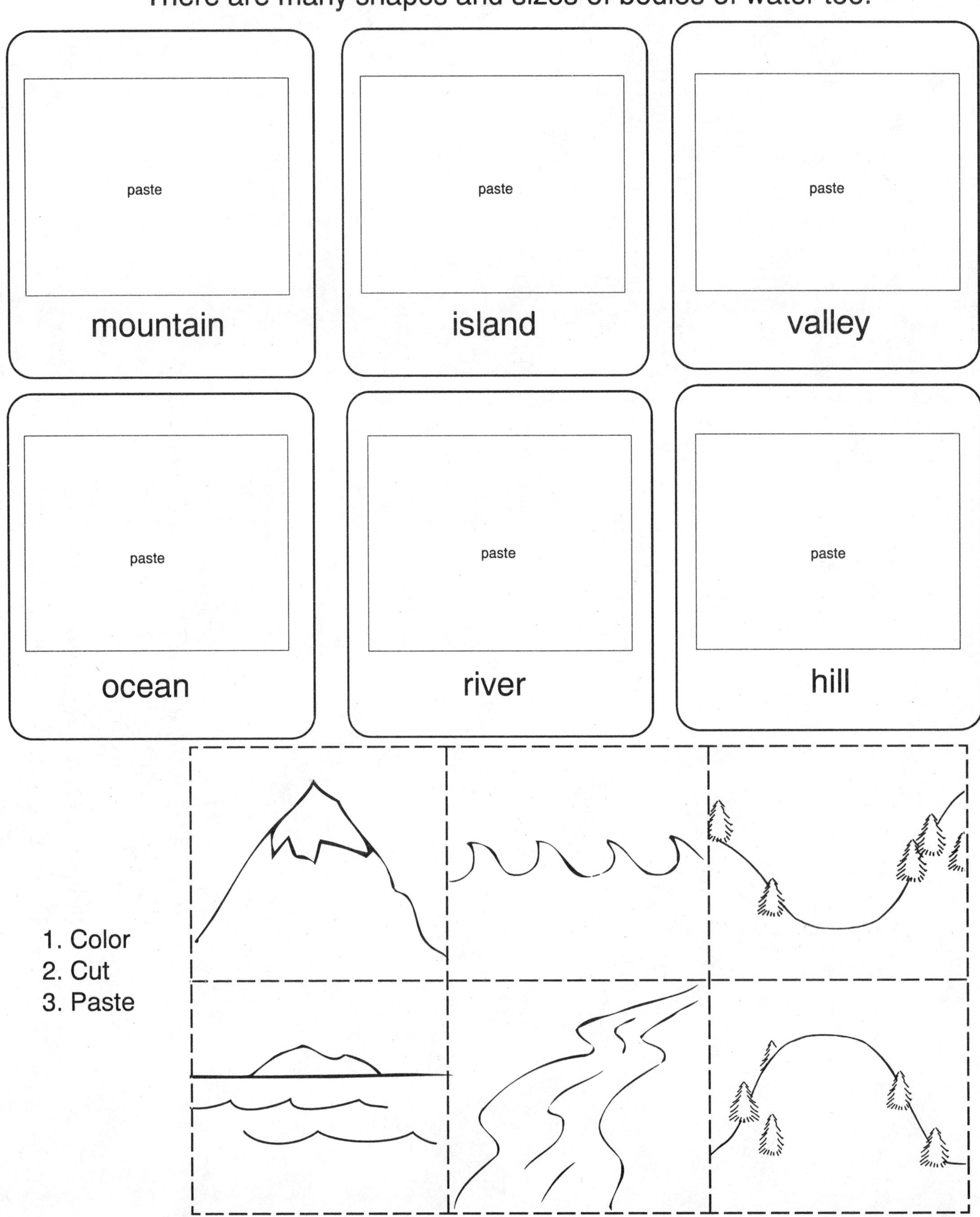

1. Color
2. Cut
3. Paste

Mountains

What do you call those parts of the earth that stand above the rest of the land?

The tallest ones are mountains.

The middle-sized ones are hills.

The shortest ones are foothills.

Have you ever tried to climb a mountain or a hill? yes ____ no ____

Note: You may want to explain that icebergs break off of glaciers and float in the sea.

Glaciers and Volcanos

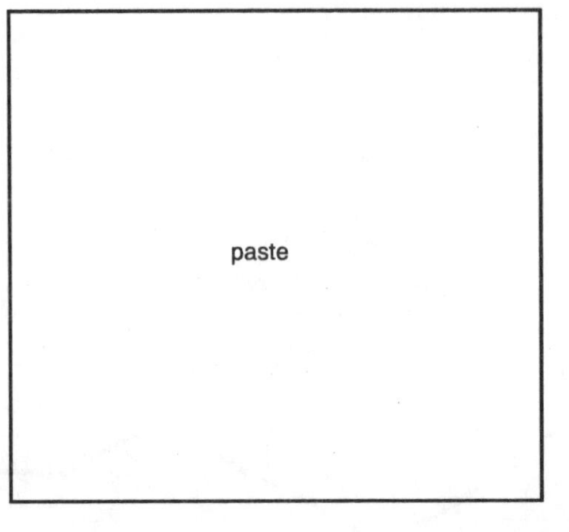

It is cold at the north and south poles all of the time. You see rivers of ice moving down the mountain sides.

These are called glaciers. Big hunks of ice break off these glaciers. The hunks float out to sea. They are called icebergs.

paste

You might see smoke and lava come out of the top of a mountain. This kind of mountain is a volcano.

Lava is hot melted rock. It pours down the sides of the volcano. When the lava cools, it is hard again.

Color, cut and paste.

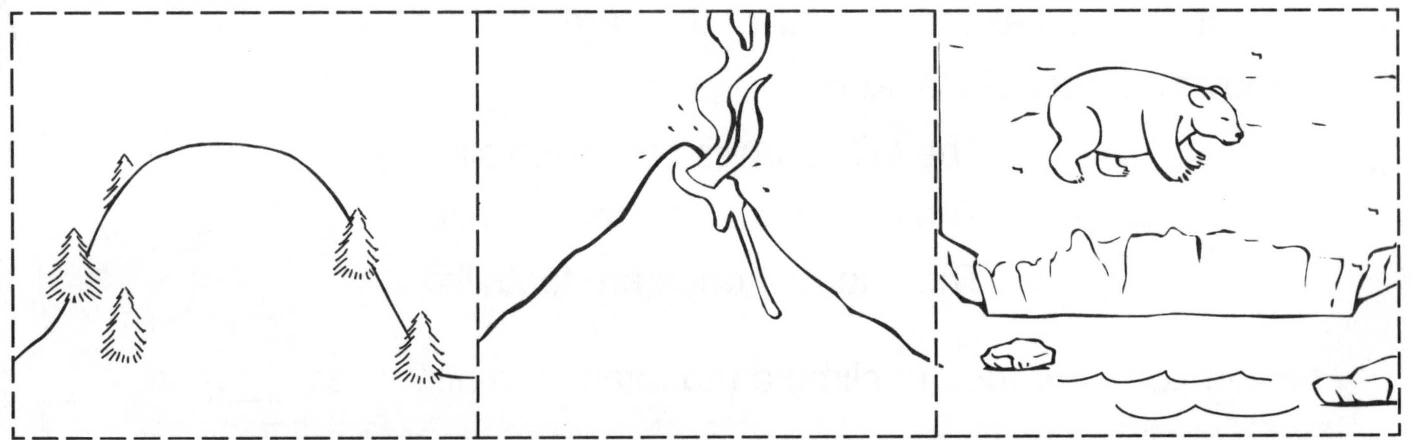

Valleys

Can you see the spaces between the mountains?

These spaces are called valleys.

Sometimes these valleys are very wide.

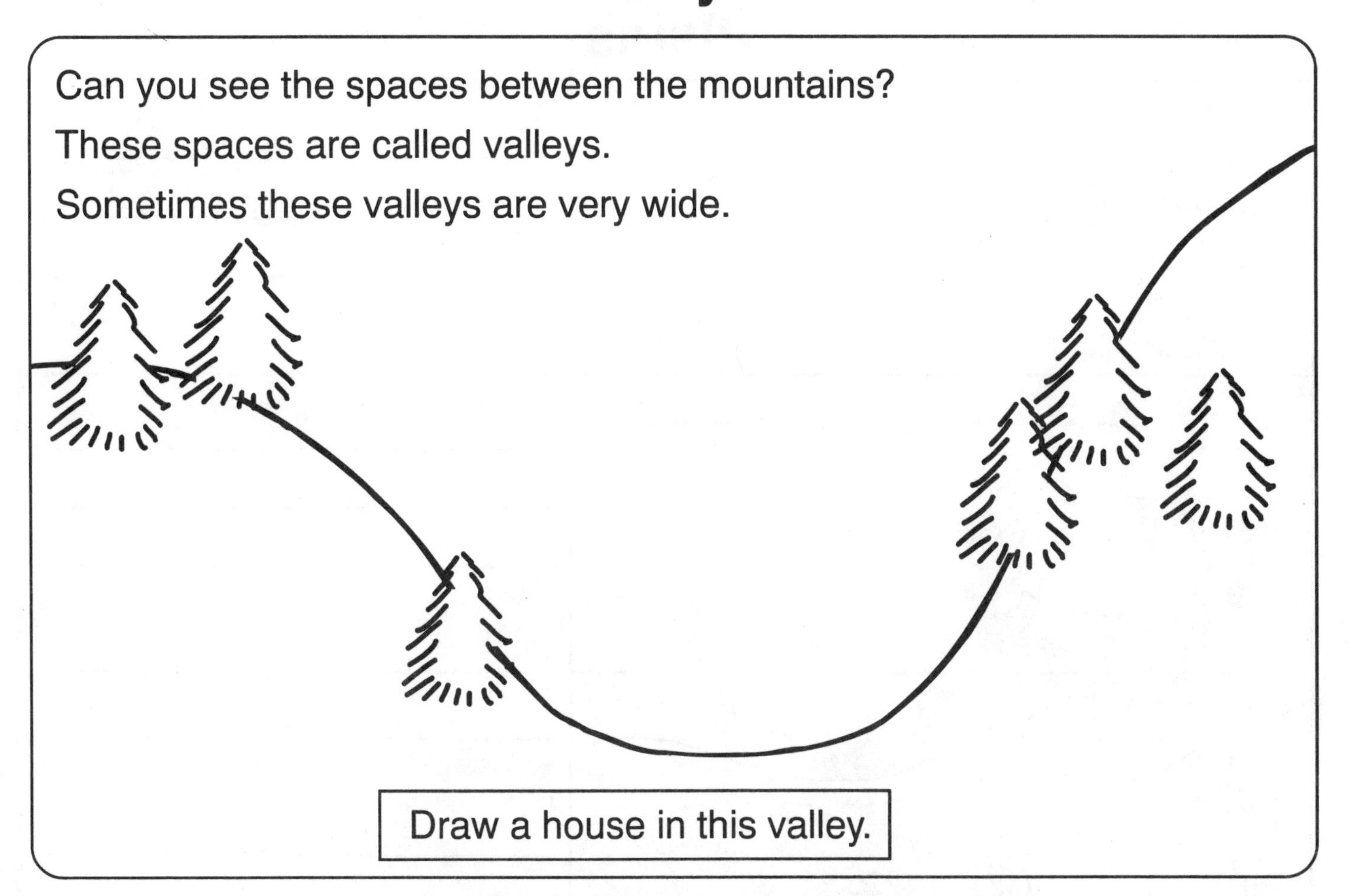

Draw a house in this valley.

Sometimes the spaces are narrow with steep sides.

Narrow valleys are called canyons.

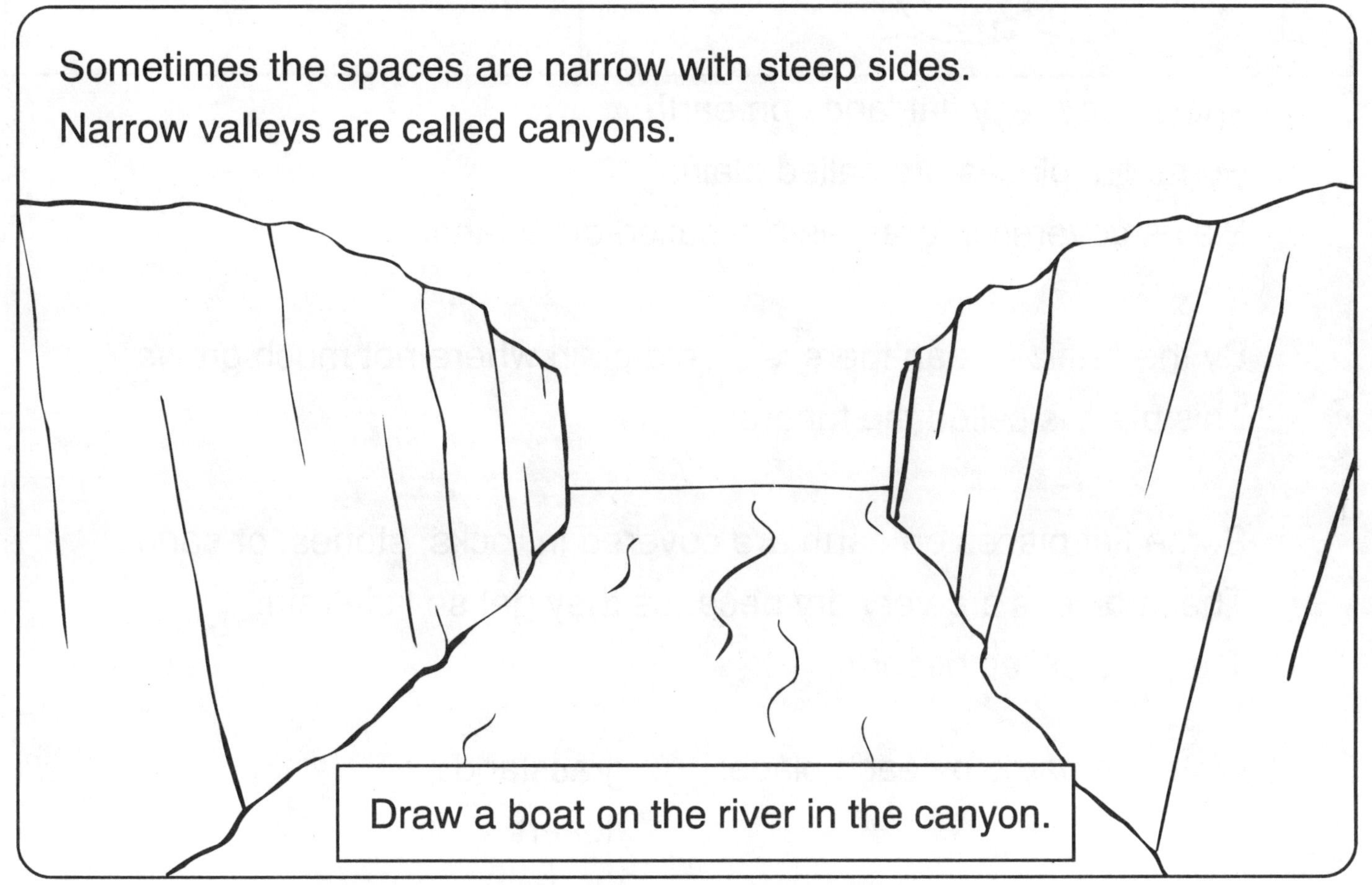

Draw a boat on the river in the canyon.

Note: You may want to share other names for plains and grasslands such as savannah, pampas, veldt, etc.
Also explain that not all deserts are flat.

Plains

There are many flat lands on earth.
Some flat places are called plains.
Plains covered in grasses are called grasslands.

By the Arctic Ocean there is a cold plain where not much grows.
This plain is called the tundra.

Some flat places on earth are covered in rocks, stones, or sand.
These places are very dry because they get so little rain.
They are called deserts.

Write the name by each place:

grassland
desert
tundra

The Seven Continents

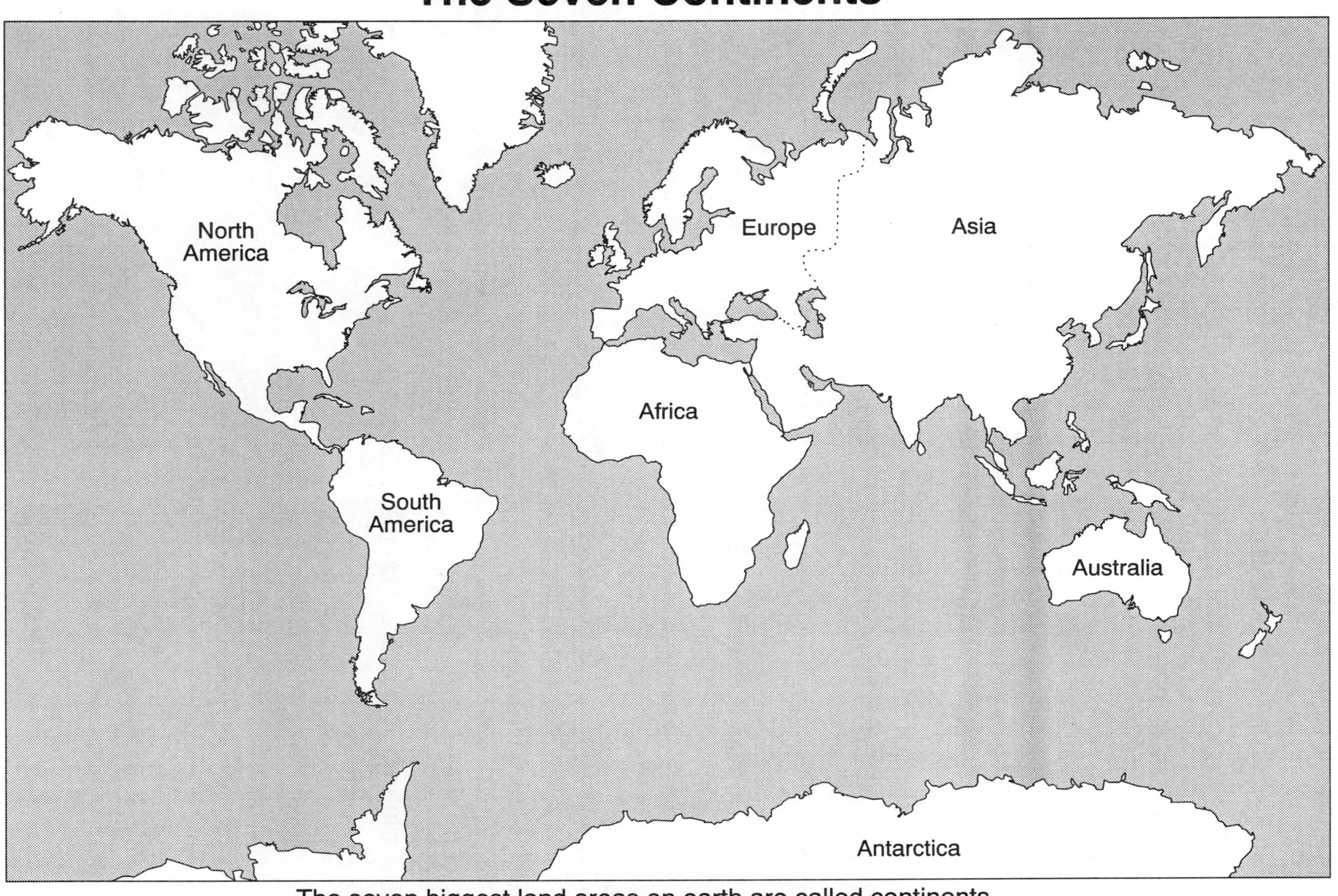

The seven biggest land areas on earth are called continents.
Color the continent on which you live with your favorite color.

Note: You may want to teach your children the word archipelago. An archipelago is a group of islands together.

Islands

Islands are much smaller than the continents.

An island is a piece of land surrounded by water.

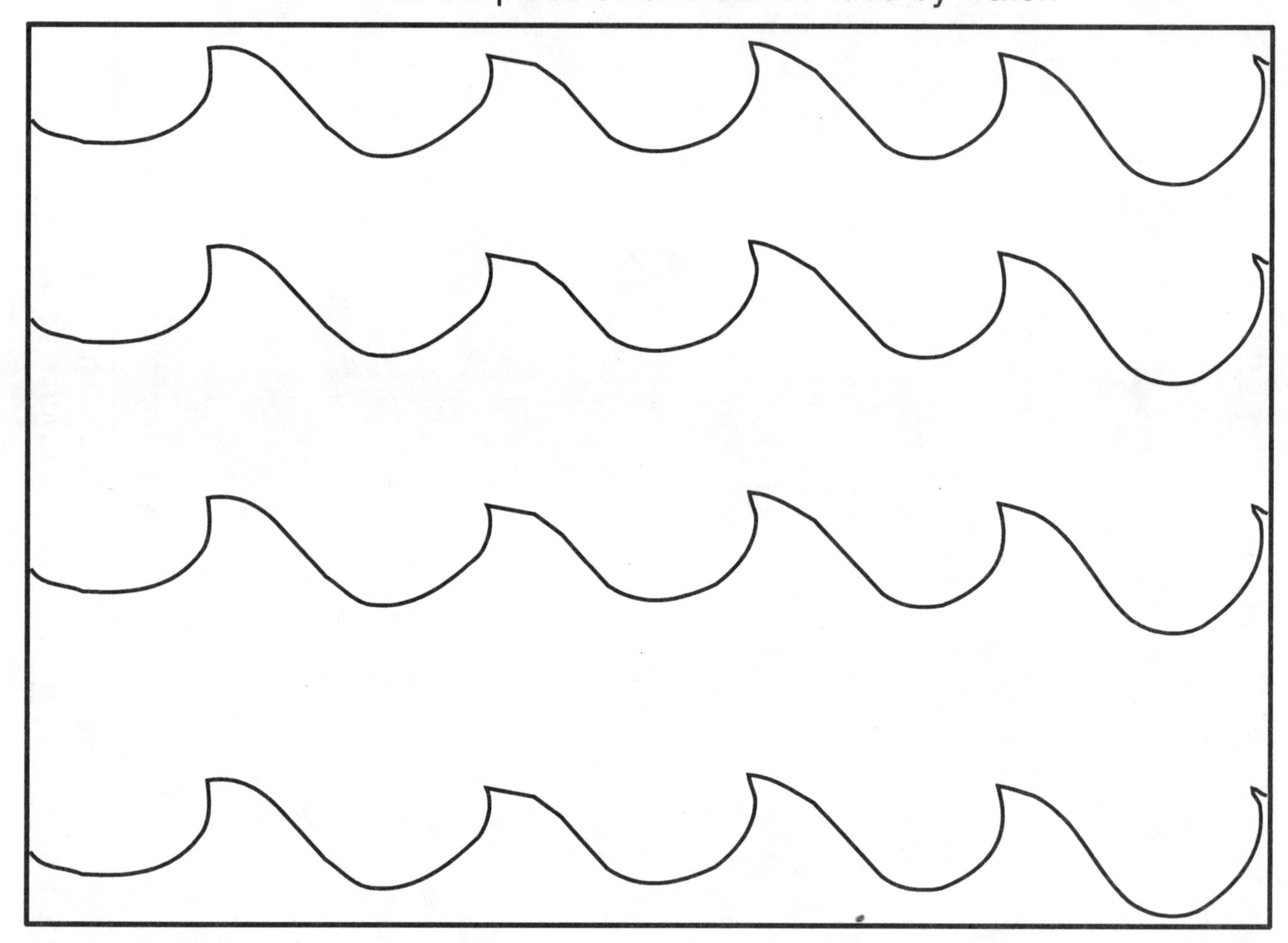

Cut out the island and paste it in the ocean.

Cut out the tree and paste it on the island.

Color the picture.

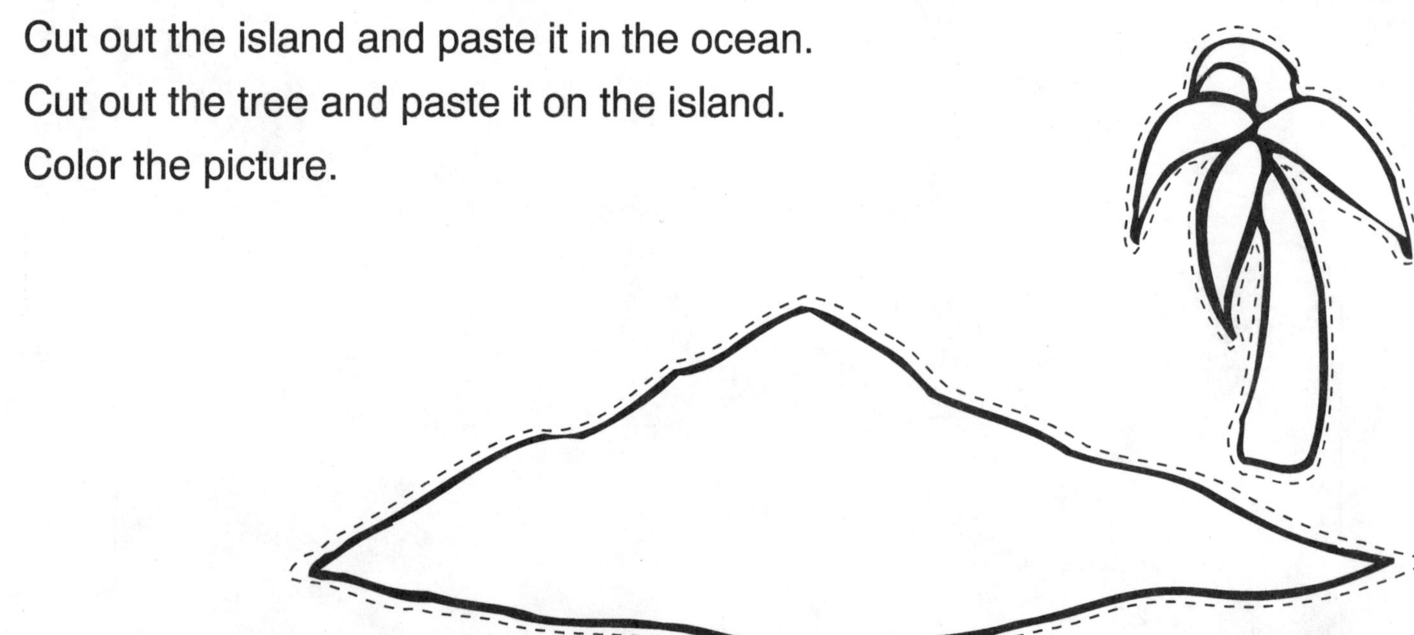

Peninsula

This is a peninsula.

It is a large piece of land that sticks far out into the water.

A peninsula has water on all but one side.

Note: You may also want to talk about gulfs, bays, fjords, etc. at this time.

Oceans

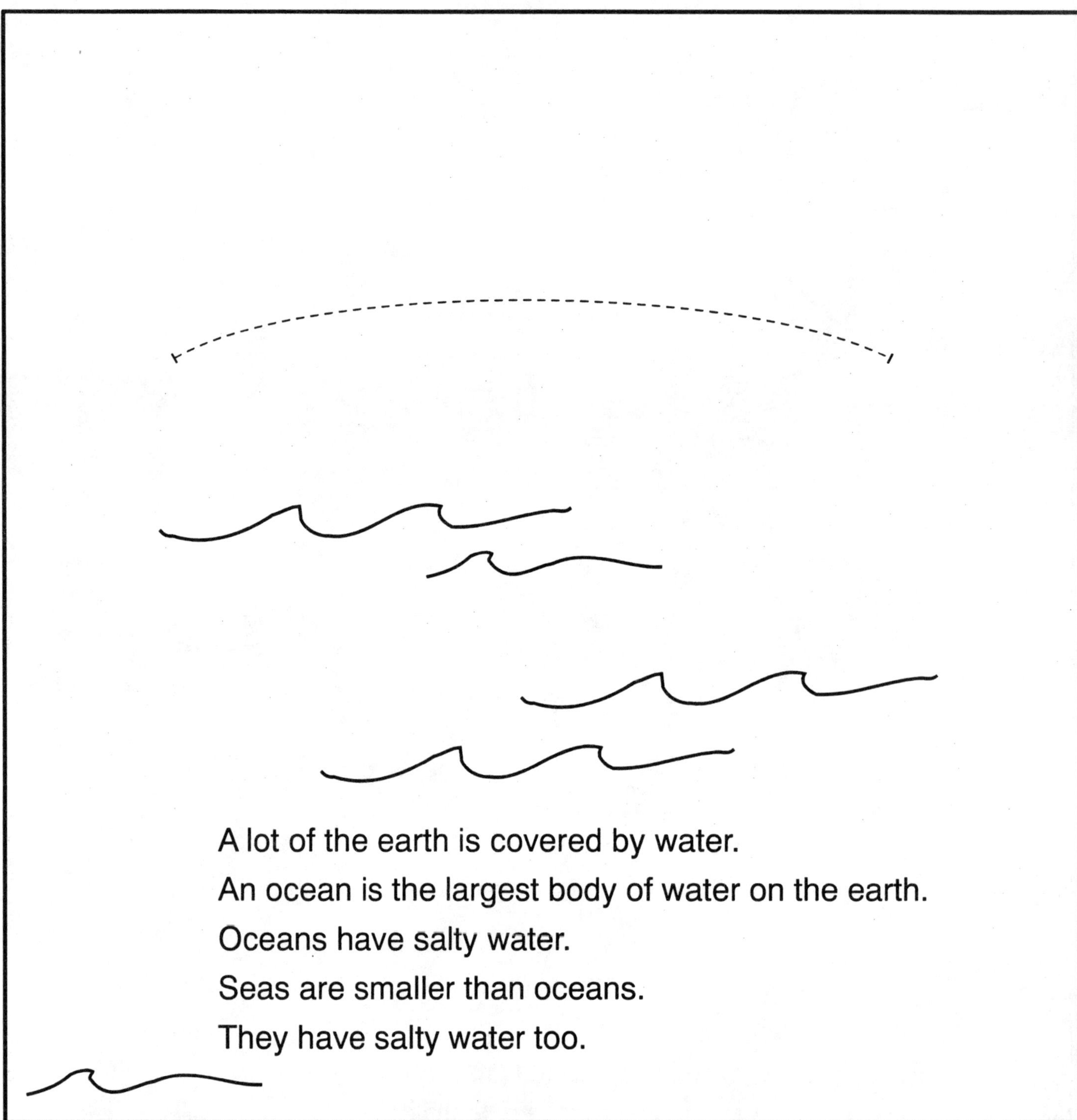

A lot of the earth is covered by water.
An ocean is the largest body of water on the earth.
Oceans have salty water.
Seas are smaller than oceans.
They have salty water too.

1. Color the ocean blue.
   Cut on the - - - - -.
2. Color and cut out the ship.
   Tape the ship to a straw.
3. Insert the straw in the cut line in the ocean.

Streams of Water

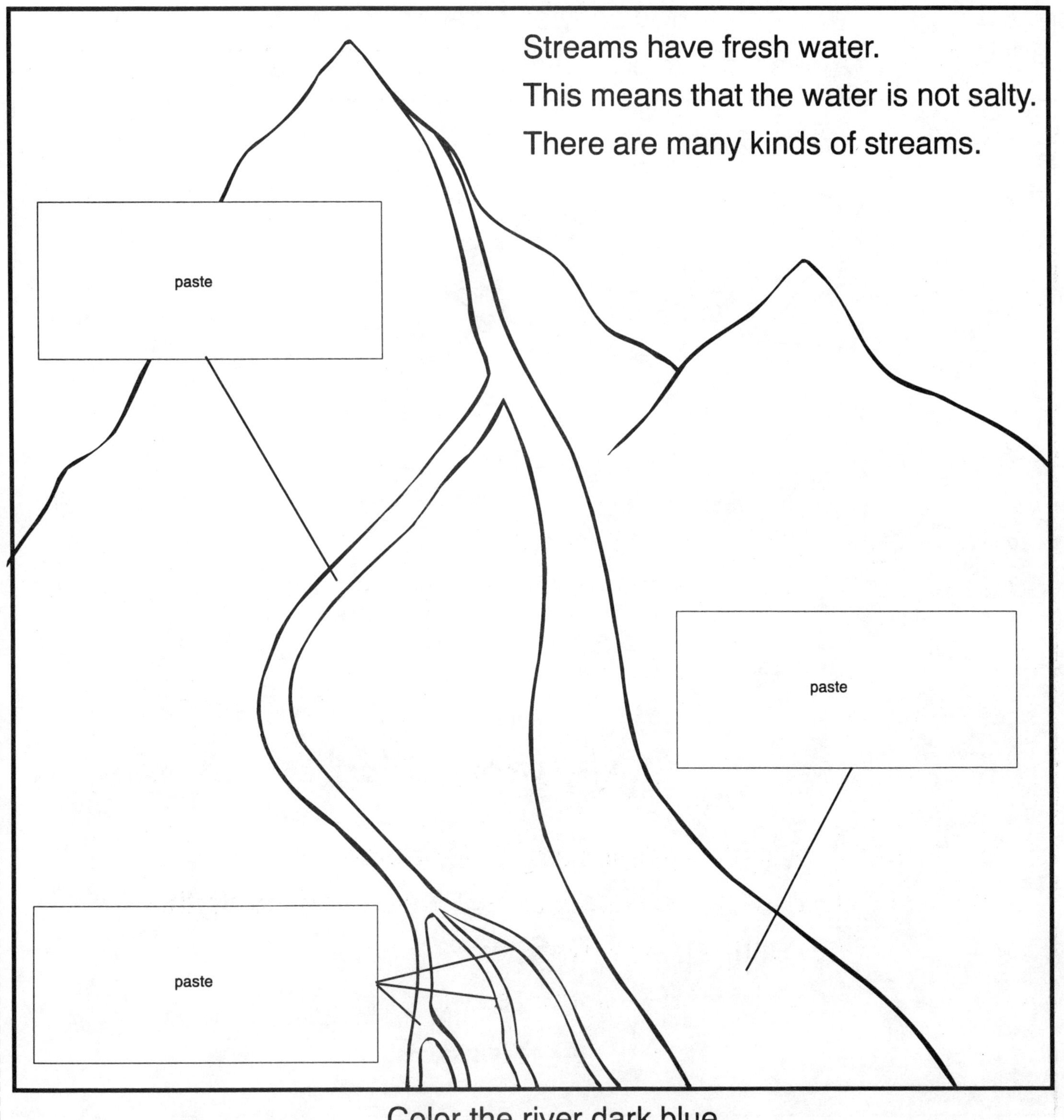

Color the river dark blue.

Color the creeks light blue.

Trace the brooks.

Rivers are long, large streams.	**Creeks** are middle-sized streams.	**Brooks** are small streams.

Lakes and Ponds

A lake is a large body of water on the land.

A pond is a smaller body of water on the land.

Paste the boat on the lake.

Paste the duck by the pond.

Is there a lake where you live? yes _____ no _____

Is there a pond where you live? yes _____ no _____

What's My Name?

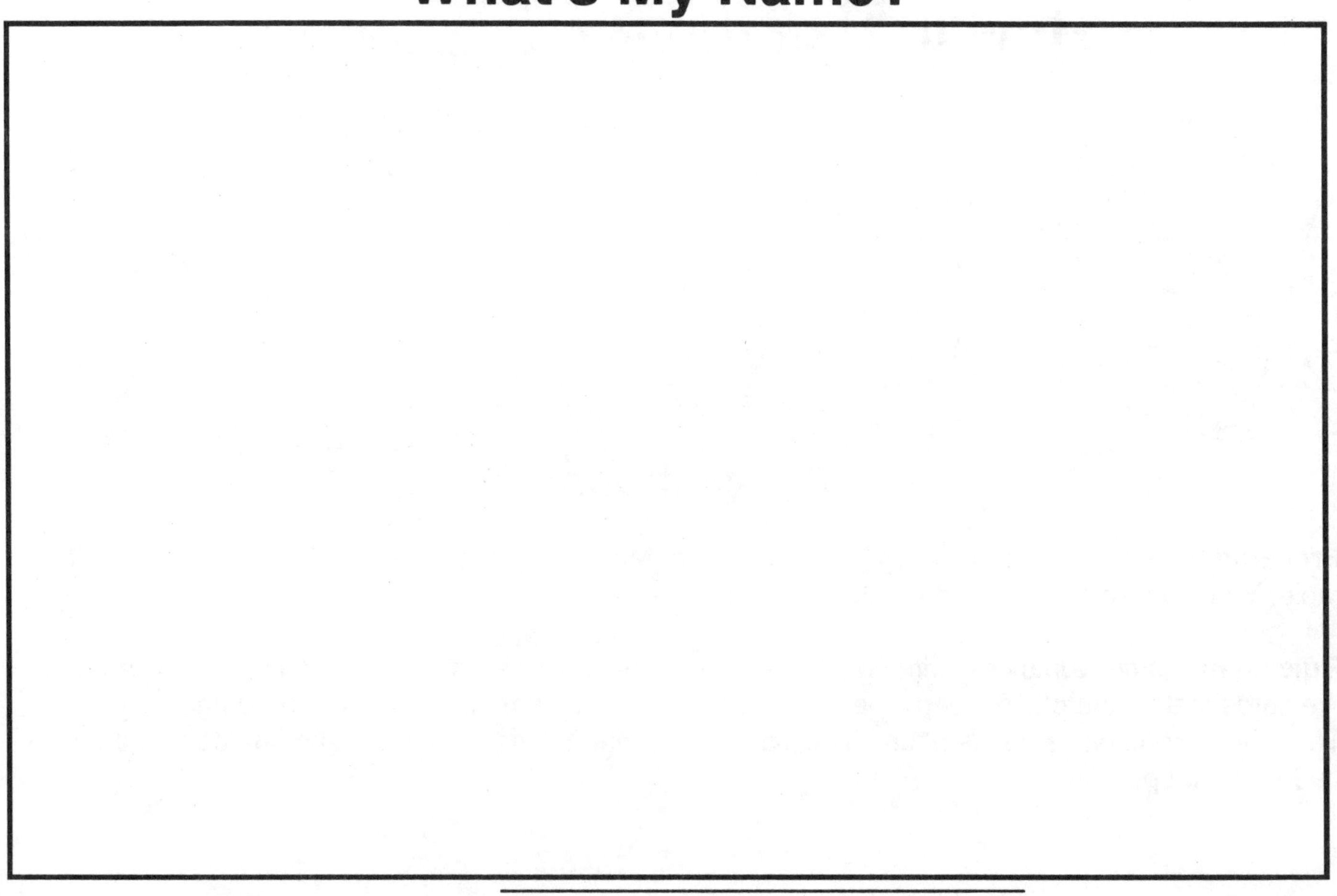

This is an _________________ .

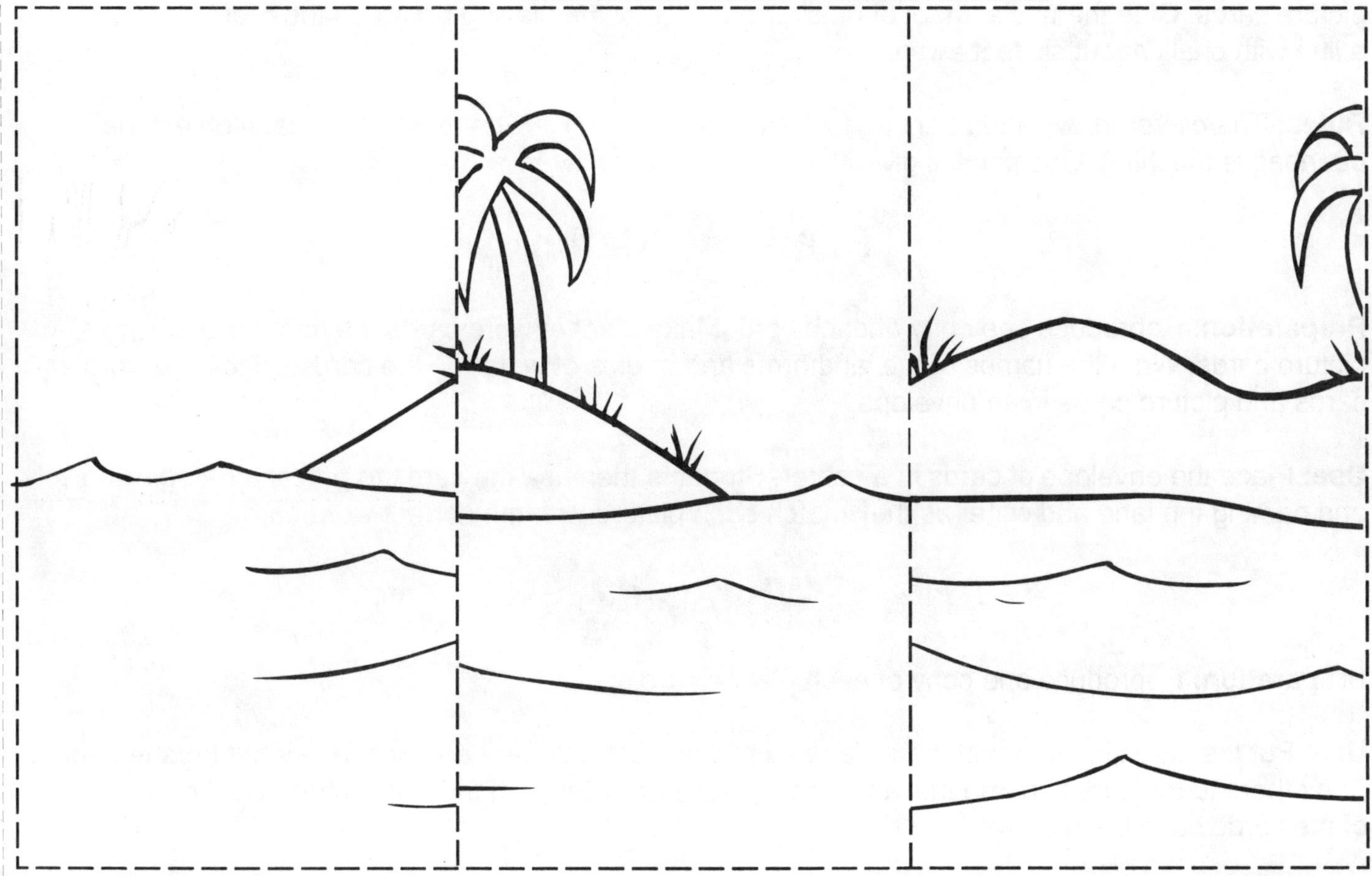

Note: Reproduce the cards on pages 15 and 16 to use with these activities. Laminate them for durability.

How to Use the Picture Cards

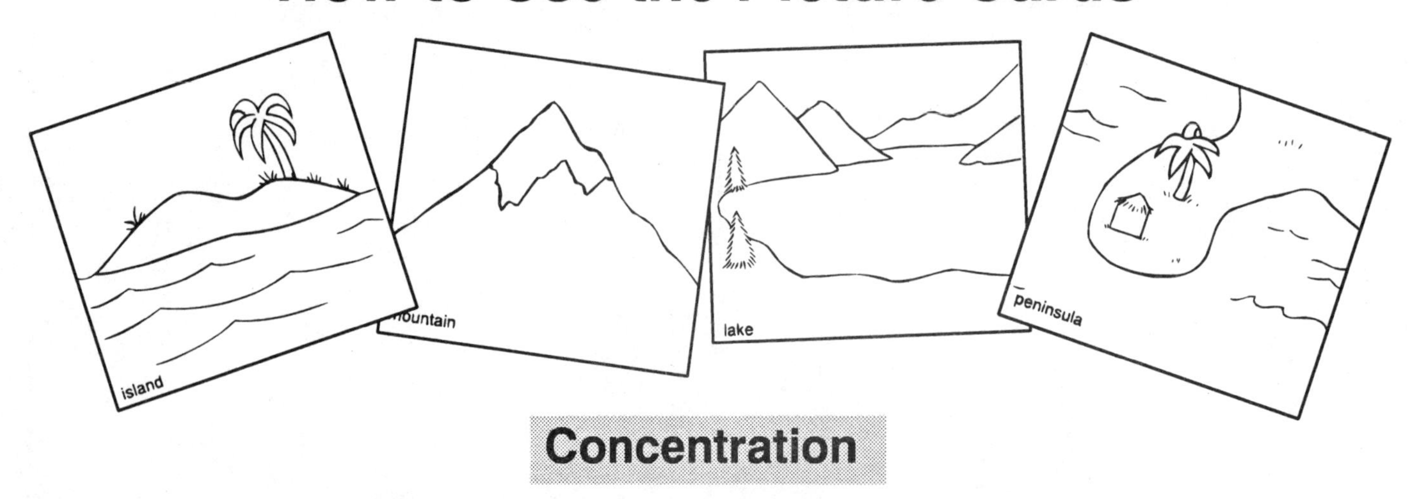

Concentration

Preparation: You will need two copies of each card. Mix the cards up. Place them face down on a table. You may want to limit the number of cards for younger players.

Rules: This game works best with only two or three players. The first player turns over two cards. If the cards match, the player keeps the cards. If the cards don't match they are turned back over. Then the second player takes a turn. The players continue taking turns until all of the cards have been picked up.

Beanbag Toss

Preparation: You will need a beanbag and a playing board. Make the playing board by enlarging the picture cards. Glue them to a sheet of butcher paper. Lay the playing board on the ground and draw a line with chalk about six feet away.

Rules: The player throws the beanbag to the playing board. He/She must name each object the beanbag is touching. One point is given for each correct answer.

Read and Match

Preparation: Reproduce one copy of each card. Make a set of word cards the same size as your picture cards. Write the names of the landforms and bodies of water on the cards. Place the word cards and picture cards in an envelope.

Use: Place the envelope of cards in a center. Students then use the cards to practice recognizing and naming the land and water as they match each picture with the correct word card.

What is it?

Preparation: Reproduce one copy of each picture card.

Use: Put the cards into a box or bag. Have a child pull out one card and describe what he/she sees. The other students try to name the land form or body of water from the description. Continue until all of the cards have been drawn.

Note: Reproduce these cards to use with the activities on page 14. Laminate them for durability.

Note: Reproduce these cards to use with the activities on page 14. Laminate them for durability.

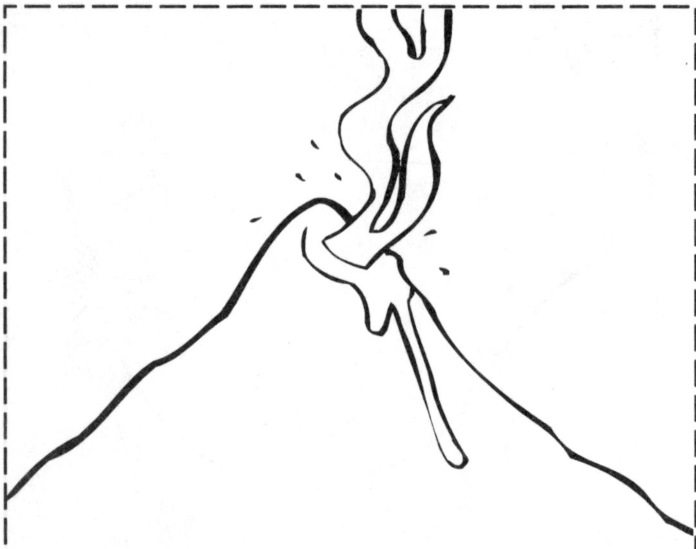

volcano

lake

island

foothills

glacier

pond